Secure Maths
Year 2

a primary maths
intervention programme

Pupil Resource Pack

Collins

William Collins' dream of knowledge for all began with the publication of his first book in 1819.
A self-educated mill worker, he not only enriched millions of lives, but also founded a flourishing publishing house. Today, staying true to this spirit, Collins books are packed with inspiration, innovation and practical expertise. They place you at the centre of a world of possibility and give you exactly what you need to explore it.

Collins. Freedom to teach.

An imprint of HarperCollins*Publishers*
The News Building
1 London Bridge Street
London
SE1 9GF

Macken House,
39/40 Mayor Street Upper,
Dublin 1
D01 C9W8
Ireland

Browse the complete Collins catalogue at
www.collins.co.uk

10 9 8 7 6 5 4 3 2 1

ISBN 978-0-00-822144-7

British Library Cataloguing in Publication Data
A catalogue record for this publication is available from the British Library.

Author Paul Hodge
Publishing manager Fiona McGlade
Editor Nina Smith
Development editor Fiona Tomlinson
Project managed by Alissa McWhinnie, QBS Learning
Copyedited by Catherine Dakin
Proofread by Cassie Fox
Answers checked by Steven Matchett
Cover design by Amparo Barrera and ink-tank and associates
Cover artwork by Amparo Barrera
Internal design by 2Hoots publishing services
Typesetting by QBS Learning
Illustrations by QBS Learning
Production by Rachel Weaver

Contents

Download Word and PDF files at collins.co.uk/pages/secure-maths-downloads

Unit 1: Count in steps of 2, 3, and 5 from 0, and in tens from any number, forward and backward

1. a) Count forwards in steps of 2 and 3 from 0 to 24.
Write the numbers.

2s: ☐ ☐ ☐ ☐ ☐ ☐ ☐

☐ ☐ ☐ ☐ ☐

3s: ☐ ☐ ☐ ☐ ☐ ☐ ☐ ☐ ☐

b) Which numbers appear in both counts? ☐

2. a) Count backwards in steps of 5 and 10 from 50 to 0.
Write the numbers.

5s: ☐ ☐ ☐ ☐ ☐ ☐

☐ ☐ ☐ ☐ ☐

10s: ☐ ☐ ☐ ☐ ☐ ☐

b) Which numbers appear in both counts? ☐

Unit 1: Count in steps of 2, 3, and 5 from 0, and in tens from any number, forward and backward

3. Number cards that were arranged in sequences of multiples of 3, multiples of 5 and multiples of 10 have got mixed up. Put them back in order.

| 10 | 20 | 9 | 0 | 0 | 30 | 30 | 12 | 20 |

| 0 | 5 | 6 | 3 | 15 | 15 | 25 | 10 |

3s

5s

10s

4. Complete the missing numbers in the grid.

10	20	30	40	50	60	70	80	90	100
110	120	130	140	150	160	170	180	190	200
210	220	230	240	250					
310	320	330	340	350					
410	420	430	440	450					

Unit 1: Count in steps of 2, 3, and 5 from 0, and in tens from any number, forward and backward

1. Continue each sequence.

a) 0, 3, 6, 9, ☐ ☐ ☐

b) 40, 50, 60, 70, ☐ ☐ ☐

c) 0, 2, 4, 6, ☐ ☐ ☐

d) 0, 5, 10, 15, ☐ ☐ ☐

2. Fill in the missing numbers.

a)

0	5		15	20		30		40	45

b)

0	2	4		8			14	16	

c)

90			120	130	140		160		180

d)

27			18		12			3	0

Unit 1: Count in steps of 2, 3, and 5 from 0, and in tens from any number, forward and backward

3. Complete the table by writing **true** or **false** in the final column.

Start number	Count in:	I say the number:	True/False
0	3s	34	
0	5s	75	
80	10s	130	
0	2s	53	

Unit 2: Recognise the place value of each digit in a two-digit number (10s and 1s)

1. Choose two cards to make the numbers below.

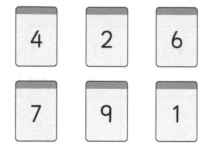

a) A number with 2 tens that is between 22 and 25

b) A number with 4 ones that is between 30 and 50

c) A number with 7 tens that is greater than 76

d) A number with 6 ones that is less than 20

2. Complete the table.

Tens	Ones	Number
10 10	1 1 1	23
10 10 10 10	1 1 1 1 1 1	
10 10 10		35
	1 1 1 1	84
		57

3. Use the number cards to make these two-digit numbers.

a) A number with 6 ones

b) A number with 8 tens

c) The largest number possible

d) The smallest number possible

e) The largest odd number

f) The smallest even number

3   8   9

4   6   1

Unit 2: Recognise the place value of each digit in a two-digit number (10s and 1s)

1. Write the missing numbers in the boxes.

a) In the number 39, there are ☐ sets of 10 and ☐ ones.

b) The number that is 6 groups of 10 is ☐.

c) The number 84 shows the digit ☐ in the tens place, and the digit ☐ in the ones place.

d) In the number 76, the value of the digit 7 is ☐ and the value of the 6 digit is ☐.

2. Work out the 'secret' two-digit numbers.

a) 'My number has 4 tens and 6 ones. What is my number?' ☐

b) 'My number has the least possible number of ones and the greatest possible number of tens. What is my number?' ☐

c) 'My number has a ones digit that is 2 more than the tens digit.' List the possible answers. (Clue: There are seven possible numbers.)

☐ ☐ ☐ ☐ ☐ ☐ ☐

d) 'My number has a tens digit that is 4 less than the ones digit.' List the possible answers. (Clue: There are five possible numbers.)

☐ ☐ ☐ ☐ ☐

3. Two children are playing a number game.

Tom says, 'My number has 3 tens and 8 ones.'

Mia says, 'My number has 3 more tens than Tom's number, but the same amount of ones.'

What are the children's numbers?

Tom: ☐   Mia: ☐

Unit 3: Identify, represent and estimate numbers using different representations, including the number line

1. Cross out the base ten blocks that **do not** correctly show the numbers.

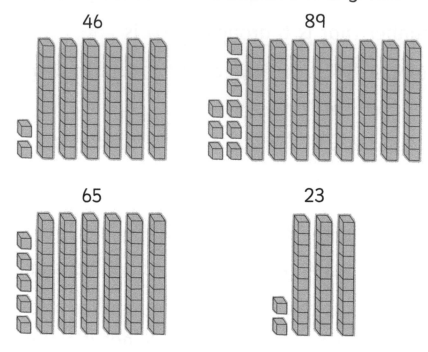

46   89

65   23

2. Draw the base ten blocks you would use to represent each number.

33		55		70		89	
T	O	T	O	T	O	T	O

3. Write the numbers in the correct places on the number lines.

a)   49   35   53   42

30   40   50   60

b)   81   92   76   98

70   80   90   100

Unit 3: Identify, represent and estimate numbers using different representations, including the number line

1. Complete the table of different number forms.

Numeral	Words	Base ten (blocks)
52	fifty-two	5 tens, 2 ones
17		1 ten, 7 ones
74	seventy-four	
28		
	thirty-nine	
		9 tens, 6 ones
	forty-one	
		5 tens

2. Write each number in expanded form. The first one has been done for you.

   a) 34 = 30 + 4   b) 67 = ☐ + ☐   c) 91 = ☐ + ☐

   d) 79 = ☐ + ☐   e) 12 = ☐ + ☐   f) 88 = ☐ + ☐

3. Grace has piles of coins made up of 10 pence and 1 pence coins.

   a) Complete the table by writing the number of coins that make up each amount.

Amount	Number of 10p coins	Number of 1p coins
13p		
62p		
38p		
45p		

   b) In the pile, there are five more 10p coins than 1p coins. The total amount is less than one pound. How much is the pile of coins worth? Write four possible answers.

   ☐ p   ☐ p   ☐ p   ☐ p

Unit 4: Compare and order numbers from 0 up to 100; use <, > and = signs

1. Use the number line to order the numbers and complete the statements.

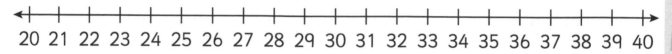

a) 32, 23

[ ] is greater than [ ]   [ ] is less than [ ]

[ ] > [ ]

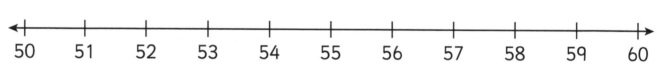

b) 57, 59

[ ] is greater than [ ]   [ ] is less than [ ]

[ ] > [ ]

2. Model both numbers in base ten blocks and compare them. Complete the statements.

a) 78, 87

[ ] is greater than [ ]   [ ] is less than [ ]

[ ] > [ ]

b) 65, 61

[ ] is greater than [ ]   [ ] is less than [ ]

[ ] > [ ]

Unit 4: Compare and order numbers from 0 up to 100; use <, > and = signs

3. Roll a dice or use a spinner to create a two-digit number. Repeat to create another two-digit number. Compare the numbers, using the symbols <, > or = in these boxes.

4. Use a number line to order the numbers, from smallest to greatest.

a) 42, 26, 51, 22

Order:

b) 28, 33, 24, 31

Order:

Unit 4: Compare and order numbers from 0 up to 100; use <, > and = signs

1. Use <, > and = signs to make these number sentences correct.

a) 72 ☐ 7 tens and 3 ones   **b)** 6 tens and 6 ones ☐ 65

c) 8 tens and seven ones ☐ 78 **d)** 59 ☐ 9 tens and 5 ones

e) 80 ☐ 7 tens and 9 ones   **f)** 3 tens and 5 ones ☐ 53

2. Write all the possible whole numbers in the boxes to make each statement true.

a) 46 < ☐ < 49   **b)** 78 < ☐ < 81

c) 32 > ☐ > 28   **d)** 100 > ☐ > 96

3. Kiera is thinking of a two-digit number. The digits of the number add up to 4. None of the digits are 0. Write a list of all the possible numbers Kiera could be thinking of then rewrite the list in order, from smallest to largest.

Possible numbers:

Order: ☐

Unit 4: Compare and order numbers from 0 up to 100; use <, > and = signs

4. Order each set of numbers, from smallest to greatest.

a) 73, 17, 37, 28

Order:

b) 34, 55, 43, 52

Order:

c) 86, 69, 84, 68, 80

Order:

d) 41, 53, 48, 52, 46

Order:

Unit 5: Read and write numbers to at least 100 in numerals and in words

1. Spin a 0–9 spinner or roll a dice to create two-digit numbers. Write the numbers in numerals and in words. Repeat for four more numbers.

Numerals	Words

2. Each set of arrow cards has been used to build a number. Write the number, both as a numeral and in words.

a) 60 ▷ sixty-three ___

3 ▷ → 6 3

b) 40 ▷ ___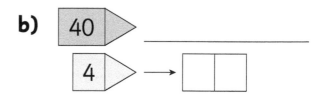

4 ▷ → ☐ ☐

c) 70 ▷ ___

1 ▷ → ☐ ☐

d) 50 ▷ ___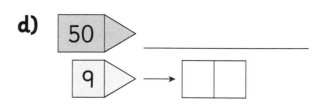

9 ▷ → ☐ ☐

e) 80 ▷ ___

8 ▷ → ☐ ☐

f) 10 ▷ ___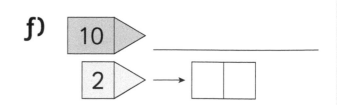

2 ▷ → ☐ ☐

Unit 5: Read and write numbers to at least 100 in numerals and in words

1. Write each number as a numeral.

Number (words)	Number (numeral)
thirty-eight	
seventeen	
eighty-one	
sixty-three	
ninety	
seventy-seven	

2. Write each number in words.

Number (numeral)	Number (words)
40	
68	
13	
96	
55	
14	

3. A 100 square has been cut into random pieces. Write the missing numbers in numerals and words.

27
twenty-seven

37
thirty-seven

63
sixty-three

74
seventy-four

17

Unit 6: Use place value and number facts to solve problems

1. Some of the runners have forgotten their numbers! Complete the missing numbers.

a)

Missing number: ☐

b)

Missing number: ☐

2. Some of these runners are standing in the wrong order. Write the numbers in the correct order.

a)

Correct order:

☐ ☐ ☐ ☐ ☐

b)

Correct order:

☐ ☐ ☐ ☐ ☐

Unit 6: Use place value and number facts to solve problems

1. Complete the next three amounts in each sequence.

   a) £42, £43, £44, _____, _____, _____

   b) 73 cm, 72 cm, 71 cm, _____, _____, _____

   c) 62 days, 64 days, 66 days, _____, _____, _____

   d) 19 hours, 16 hours, 13 hours, _____, _____, _____

2. Guess each two-digit number from the clues.

   a) My number:
   - is less than 50
   - has a tens digit that is four more than its ones digit.

   My number is ☐.

   b) My number:
   - is greater than 70
   - has a ones digit that is two more than its tens digit.

   My number is ☐.

   c) My number:
   - is between 32 and 46
   - has a tens digit that is three less than its ones digit.

   My number is ☐.

   d) My number:
   - is between 40 and 56
   - has a ones digit that is four more than its tens digit.

   My number is ☐.

3. Order each set of numbers, from least to greatest, and identify the number in the middle of the sequence.

Middle number

a)

14	24	42	23	13		

b)

45	42	54	50	44		

c)

78	67	87	57	75	60	77	

d)

98	86	89	81	90	91	93	

Unit 7: Solve problems with addition and subtraction: using concrete objects and pictorial representations, including those involving numbers, quantities and measures; applying their increasing knowledge of mental and written methods

1. Work out these addition calculations. Use the base ten blocks to help you.

   a) 26 centimetres + 13 centimetres

   = ☐ cm

   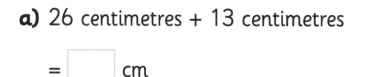

   b) 32 litres + 24 litres = ☐ l

   c) 45p + 31p = ☐ p

   d) 74 grams + 25 grams = ☐ g

   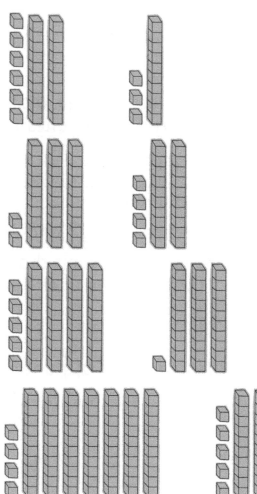

2. Work out these subtraction calculations. Use the base ten blocks to help you.

   a) 47 centimetres – 22 centimetres

   = ☐ cm

   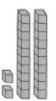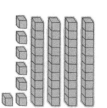

Unit 7: Solve problems with addition and subtraction: using concrete objects and pictorial representations, including those involving numbers, quantities and measures; applying their increasing knowledge of mental and written methods

b) 68 litres – 45 litres = ☐ l

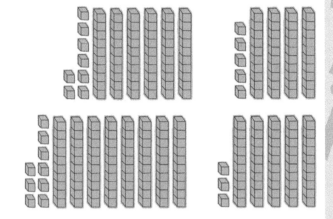

c) 89p – 53p = ☐ p

3. Work out these calculations. Use the number lines to help you.

a) 36 + 23 = ☐

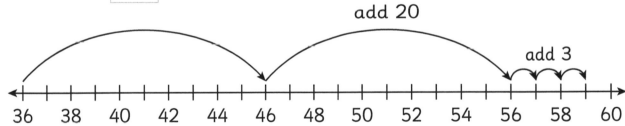

add 20

add 3

36 38 40 42 44 46 48 50 52 54 56 58 60

b) 62 + 37 = ☐

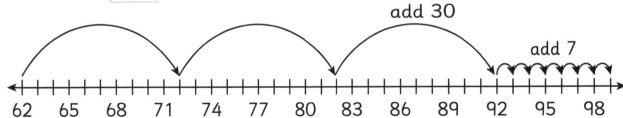

add 30

add 7

62 65 68 71 74 77 80 83 86 89 92 95 98

c) 57 – 44 = ☐

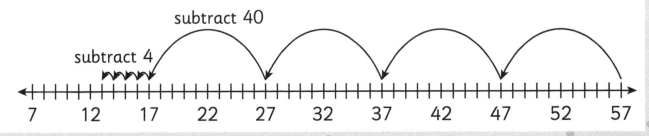

subtract 40

subtract 4

7   12   17   22   27   32   37   42   47   52   57

Unit 7: Solve problems with addition and subtraction: using concrete objects and pictorial representations, including those involving numbers, quantities and measures; applying their increasing knowledge of mental and written methods

1. Work out these addition calculations.

   a) 53 centimetres + 22 centimetres = ☐ cm

   b) 48 litres + 31 litres = ☐ l

   c) 67p + 12p = ☐ p

   d) 72 grams + 27 grams = ☐ g

2. Work out these subtraction calculations.

   a) 59 centimetres − 33 centimetres = ☐ cm

   b) 48 litres − 27 litres = ☐ l

   c) 93p − 42p = ☐ p

   d) 77 grams − 55 grams = ☐ g

3. Work out these calculations. Draw a number line on a separate piece of paper to help you.

   a) 42 + 26 = ☐   **b)** 75 + 24 = ☐

   c) 94 − 73 = ☐   **d)** 68 − 14 = ☐

Unit 7: Solve problems with addition and subtraction: using concrete objects and pictorial representations, including those involving numbers, quantities and measures; applying their increasing knowledge of mental and written methods

4. Work out the answers to these word problems.

a) 47 frogs live in a pond. 34 frogs are sitting on lily pads. How many frogs are not on the lily pads?

[   ] frogs

b) 33 doors on the first floor of a block of flats are red. 26 doors on the next floor are also red. If the block has just two floors, how many flats altogether have red doors?

[   ] flats

c) 66 owls are sitting on the branches of a tree. Later, 35 of them fly away. How many owls are left?

[   ] owls

d) 72 people are sitting on a bus. At the next stop, 23 people get on. How many people in total are now on the bus?

[   ] people

Unit 8: Recall and use addition and subtraction facts to 20 fluently, and derive and use related facts up to 100

1. Each bead has a value of 10. Write the number statement shown by each set of beads. The first one has been done for you.

a)

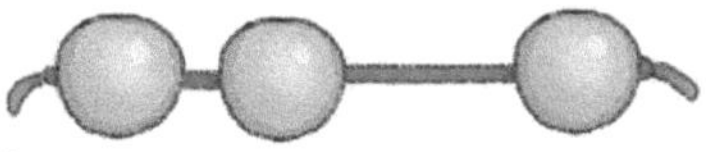

$$20 + 10 = 30$$

b)

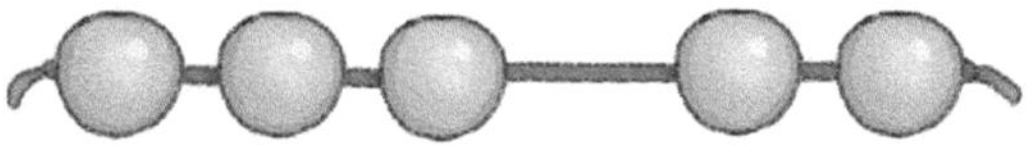

☐ + ☐ = ☐

c)

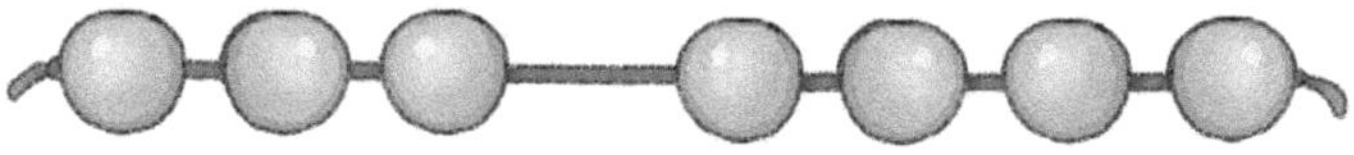

☐ + ☐ = ☐

d)

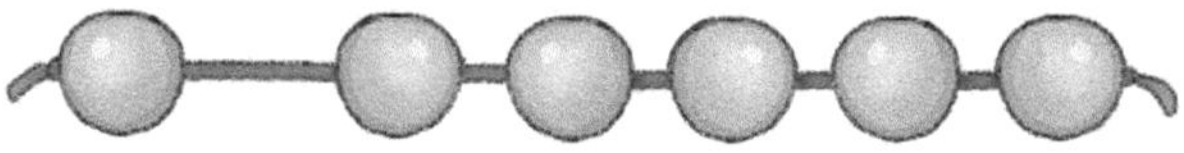

☐ + ☐ = ☐

e)

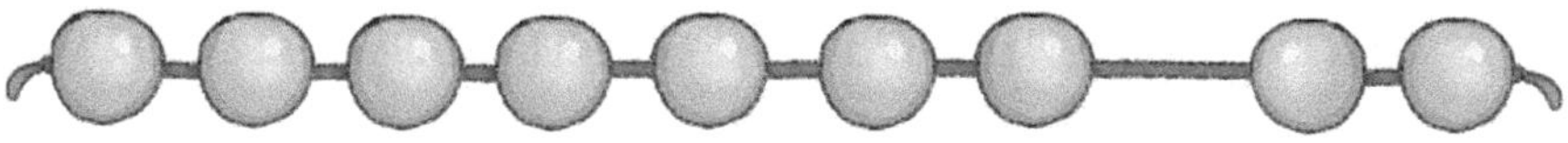

☐ + ☐ = ☐

f)

☐ + ☐ = ☐

Unit 8: Recall and use addition and subtraction facts to 20 fluently, and derive and use related facts up to 100

2. Each sock on the washing line has a value of 10. Write the number statement shown by each set of socks. The first one has been done for you.

a)

50 – 40 = ☐

b)

☐ – ☐ = ☐

c)

☐ – ☐ = ☐

d)

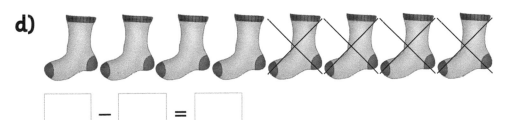

☐ – ☐ = ☐

e)

☐ – ☐ = ☐

f)

☐ – ☐ = ☐

Unit 8: Recall and use addition and subtraction facts to 20 fluently, and derive and use related facts up to 100

1. Complete the addition statements.

a) 4 + 3 =

b) 7 + 4 =

c) 5 + 9 =

d) 12 + 2 =

e) 4 + 14 =

f) 18 + 2 =

g) 8 + 11 =

h) 13 + 3 =

i) 4 + 16 =

2. Complete the subtraction statements.

a) 8 – 5 =

b) 9 – 3 =

c) 7 – 6 =

d) 15 – 6 =

e) 13 – 8 =

f) 14 – 7 =

g) 18 – 9 =

h) 20 – 12 =

i) 19 – 13 =

Unit 8: Recall and use addition and subtraction facts to 20 fluently, and derive and use related facts up to 100

3. Use one fact you know to complete one you may not know.

a) 6 + 2 =

60 + 20 =

b) 3 + 5 =

30 + 50 =

c) 5 + 4 =

50 + 40 =

d) 8 + 2 =

80 + 20 =

e) 1 + 7 =

10 + 70 =

f) 3 + 6 =

30 + 60 =

g) 7 – 2 =

70 – 20 =

h) 8 – 4 =

80 – 40 =

i) 9 – 7 =

90 – 70 =

j) 6 – 5 =

60 – 50 =

k) 9 – 8 =

90 – 80 =

l) 10 – 4 =

100 – 40 =

4. Complete the missing numbers.

a) 40 + 20 =

b) 50 + = 60

c) + 40 = 70

d) 70 – 30 =

e) 80 – = 30

f) – 20 = 70

Unit 9: Add and subtract numbers using concrete objects, pictorial representations, and mentally, including: a two-digit number and 1s; a two-digit number and 10s; 2 two-digit numbers; adding 3 one-digit numbers

1. a) Make two-digit numbers by taking a tens card from the top row and a ones card from the bottom row.

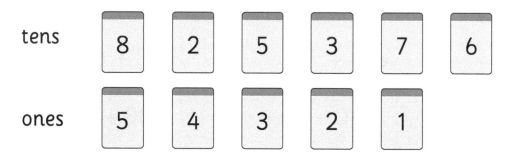

tens: 8 2 5 3 7 6

ones: 5 4 3 2 1

Add the two-digit numbers to another card from the bottom row. Write the addition as a number statement. Use a number line or base ten blocks to help you.

i) ☐ + ☐ = ☐   **ii)** ☐ + ☐ = ☐

iii) ☐ + ☐ = ☐   **iv)** ☐ + ☐ = ☐

b) Choose cards to make and solve four two-digit, subtract one-digit number statements. The ones digit in the two-digit number must be greater than the digit in the one-digit number.

i) ☐ – ☐ = ☐   **ii)** ☐ – ☐ = ☐

iii) ☐ – ☐ = ☐   **iv)** ☐ – ☐ = ☐

Unit 9: Add and subtract numbers using concrete objects, pictorial representations, and mentally, including: a two-digit number and 1s; a two-digit number and 10s; 2 two-digit numbers; adding 3 one-digit numbers

2. a) Make two-digit numbers by choosing two cards from the top row.

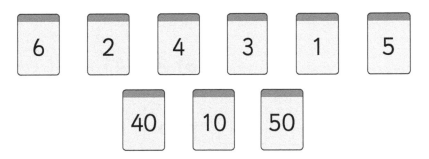

Add each two-digit number to a card from the bottom row. Write the addition as a number statement. Use a number line or base ten blocks to help you.

i) ☐ + ☐ = ☐   **ii)** ☐ + ☐ = ☐

iii) ☐ + ☐ = ☐   **iv)** ☐ + ☐ = ☐

b) Choose cards to make and solve four two-digit 'number subtract a multiple of ten' number statements.

i) ☐ – ☐ = ☐   **ii)** ☐ – ☐ = ☐

iii) ☐ – ☐ = ☐   **iv)** ☐ – ☐ = ☐

Unit 9: Add and subtract numbers using concrete objects, pictorial representations, and mentally, including: a two-digit number and 1s; a two-digit number and 10s; 2 two-digit numbers; adding 3 one-digit numbers

1. Complete the addition statements.

a) $12 + 3 =$

b) $4 + 33 =$

c) $57 + 2 =$

d) $8 + 41 =$

e) $83 + 3 =$

f) $6 + 72 =$

g) $10 + 23 =$

h) $48 + 10 =$

i) $10 + 87 =$

j) $32 + 30 =$

k) $20 + 64 =$

l) $41 + 40 =$

2. Complete the subtraction statements.

a) $25 - 3 =$

b) $47 - 5 =$

c) $68 - 6 =$

d) $36 - 2 =$

e) $59 - 8 =$

f) $74 - 1 =$

g) $54 - 10 =$

h) $76 - 10 =$

i) $98 - 10 =$

j) $63 - 20 =$

k) $87 - 40 =$

l) $76 - 60 =$

Unit 9: Add and subtract numbers using concrete objects, pictorial representations, and mentally, including: a two-digit number and 1s; a two-digit number and 10s; 2 two-digit numbers; adding 3 one-digit numbers

3. Add the money, using your preferred method.

a) 31p + 12p = ☐ p   **b)** 23p + 54p = ☐ p

c) 42p + 34p = ☐ p   **d)** 26p + 53p = ☐ p

e) 67p + 22p = ☐ p   **f)** 16p + 72p = ☐ p

g) 17p + 24p = ☐ p   **h)** 39p + 14p = ☐ p

i) 25p + 38p = ☐ p   **j)** 64p + 28p = ☐ p

k) 47p + 36p = ☐ p   **l)** 49p + 49p = ☐ p

4. Work out the total of each row and column.

5	9	3	
6	8	8	
4	1	7	

Unit 10: Show that addition of 2 numbers can be done in any order (commutative) and subtraction of 1 number from another cannot

1. Complete each set of number sentences using the number cards. Each number card may only be used once in each sentence. The symbol ≠ means 'does not equal'.

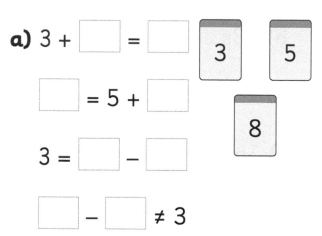

a) 3 + ☐ = ☐

☐ = 5 + ☐

3 = ☐ − ☐

☐ − ☐ ≠ 3

3   5   8

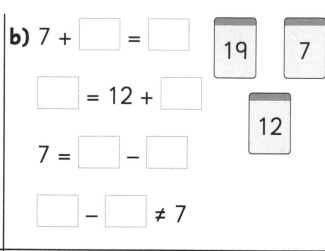

b) 7 + ☐ = ☐

☐ = 12 + ☐

7 = ☐ − ☐

☐ − ☐ ≠ 7

19   7   12

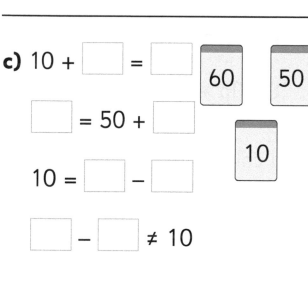

c) 10 + ☐ = ☐

☐ = 50 + ☐

10 = ☐ − ☐

☐ − ☐ ≠ 10

60   50   10

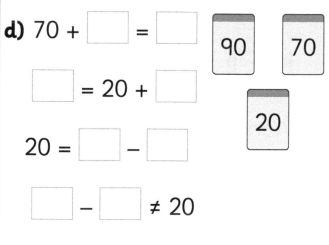

d) 70 + ☐ = ☐

☐ = 20 + ☐

20 = ☐ − ☐

☐ − ☐ ≠ 20

90   70   20

2. Use the number cards to make as many addition and subtraction sentences as you can. How many can you make?

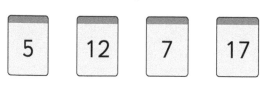

5   12   7   17

Unit 10: Show that addition of 2 numbers can be done in any order (commutative) and subtraction of 1 number from another cannot

1. Circle two of the number sentences that show addition can be done in any order.

7 + 4 = 11	16 + 7 = 23	18 = 7 + 11
9 + 7 = 16	11 + 7 = 18	16 + 9 = 15

2. Solve each addition problem. Choose the quickest method possible.

a) 4 + 73 = ☐   **b)** 6 + 91 = ☐   **c)** 5 + 84 = ☐

d) 12 + 66 = ☐   **e)** 15 + 81 = ☐   **f)** 17 + 72 = ☐

3. The number at the top of a bar model is the sum of the two numbers below it. Use the bar model below to complete four number sentences. The symbol ≠ means 'does not equal'.

☐ + ☐ = 19   ☐ + ☐ = 19

☐ − ☐ = 14   ☐ − ☐ ≠ 14

19	
14	5

4. Does each pair of calculations give the same answer? Write **yes** or **no** to complete the table.

Calculations	Same answer (yes/no)
7 − 4 and 4 − 7	
9 + 28 and 28 + 9	
45 + 54 and 54 + 45	
86 − 19 and 19 − 86	

Unit 11: Recognise and use the inverse relationship between addition and subtraction and use this to check calculations and solve missing number problems

1. Draw a bar model diagram to illustrate each number sentence.

a) 9 + 4 = 13

b) 21 = 12 + 9

c) 7 = 19 – 12

d) 28 – 15 = 13

2. Draw a number bond diagram to illustrate each number sentence.

a) 35 + 14 = 49

b) 19 = 40 – 21

3. In each row, colour the number sentence that is the **inverse operation** of the first calculation and shows the answer is correct.

a) 19 + 7 = 26   7 + 19 = 26   26 – 19 = 7   20 + 6 = 26

b) 25 – 8 = 17   7 + 19 = 26   18 + 7 = 25   8 + 17 = 25

c) 42 = 16 + 26   42 – 16 = 26   58 – 26 = 32   26 + 16 = 42

Unit 11: Recognise and use the inverse relationship between addition and subtraction and use this to check calculations and solve missing number problems

1. Complete the missing number in each bar model.

a)

18	
12	

b)

27	
	16

c)

54	
38	

d)

82	
	65

2. Write the missing numbers.

a) $4 + \boxed{} = 12$

b) $\boxed{} + 14 = 22$

c) $20 + \boxed{} = 34$

d) $\boxed{} + 17 = 41$

e) $36 + \boxed{} = 50$

f) $\boxed{} + 68 = 84$

g) $16 - \boxed{} = 11$

h) $\boxed{} + 5 = 18$

i) $24 - \boxed{} = 12$

j) $\boxed{} - 30 = 63$

k) $\boxed{} + 34 = 58$

l) $90 - \boxed{} = 58$

3. Write an inverse operation for each number sentence.

a) $5 + 14 = 19$ _____

b) $14 - 8 = 6$ _____

c) $26 + 11 = 37$ _____

d) $75 - 22 = 53$ _____

4. Use an inverse operation to check each calculation. Circle the calculations that are incorrect.

a) $40 - 7 = 33$

b) $15 + 9 = 24$

c) $34 - 16 = 12$

d) $22 + 54 = 78$

e) $68 - 32 = 36$

f) $26 + 72 = 100$

Unit 12: Recall and use multiplication and division facts for the 2, 5 and 10 multiplication tables, including recognising odd and even numbers

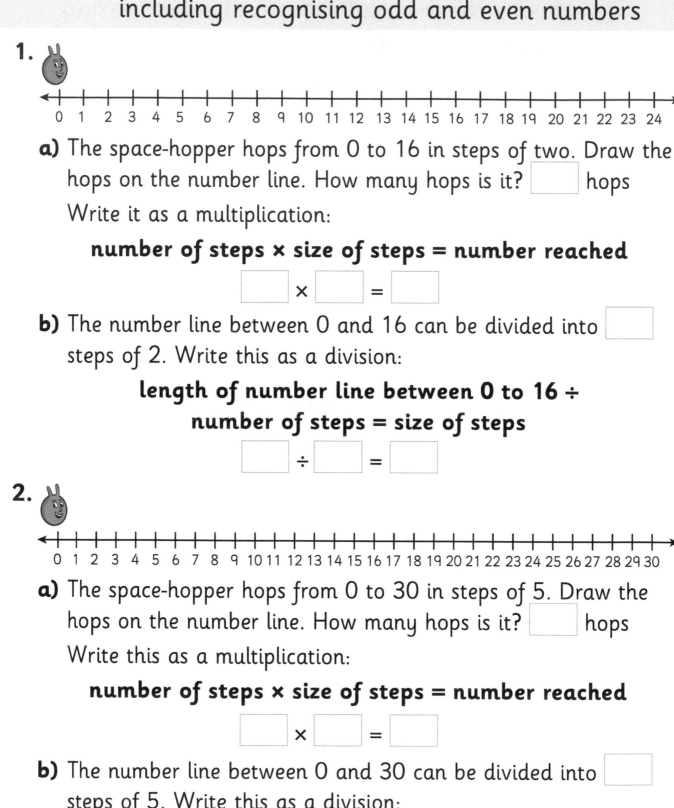

1.

0 1 2 3 4 5 6 7 8 9 10 11 12 13 14 15 16 17 18 19 20 21 22 23 24

a) The space-hopper hops from 0 to 16 in steps of two. Draw the hops on the number line. How many hops is it? ☐ hops

Write it as a multiplication:

number of steps × size of steps = number reached

☐ × ☐ = ☐

b) The number line between 0 and 16 can be divided into ☐ steps of 2. Write this as a division:

length of number line between 0 to 16 ÷ number of steps = size of steps

☐ ÷ ☐ = ☐

2.

0 1 2 3 4 5 6 7 8 9 10 11 12 13 14 15 16 17 18 19 20 21 22 23 24 25 26 27 28 29 30

a) The space-hopper hops from 0 to 30 in steps of 5. Draw the hops on the number line. How many hops is it? ☐ hops

Write this as a multiplication:

number of steps × size of steps = number reached

☐ × ☐ = ☐

b) The number line between 0 and 30 can be divided into ☐ steps of 5. Write this as a division:

length of number line between 0 to 30 ÷ number of steps = size of steps

☐ ÷ ☐ = ☐

Unit 12: Recall and use multiplication and division facts for the 2, 5 and 10 multiplication tables, including recognising odd and even numbers

1. Complete the missing numbers in each sequence.

a) 0, 2, 4, ☐, 8, 10, ☐, 14, 16, ☐, 20

b) 0, 5, 10, ☐, 20, ☐, 30, 35, ☐, 45

c) 0, 10, 20, 30, ☐, ☐, 60, 70, ☐, 90

2. Write the answer to each multiplication.

a) 6×2 = ☐

b) 10×2 = ☐

c) 12×2 = ☐

d) 5×5 = ☐

e) 7×5 = ☐

f) 11×5 = ☐

g) 4×10 = ☐

h) 8×10 = ☐

i) 12×10 = ☐

3. In each pair of circles, colour the circle that has the larger value.

a) (7×5) or (4×10)

b) (9×2) or (4×5)

c) (2×10) or (11×2)

d) (12×5) or (7×10)

4. Answer these questions.

a) A ribbon measuring 10 centimetres is divided into 5 equal pieces. What is the length of each piece? ☐ centimetres

b) I have 90p in my pocket in 10p coins. How many coins do I have? ☐ coins

c) Each section of a fence is made from 5 pieces of wood. If a fence is made from 35 pieces of wood, how many sections are there? ☐ sections

d) I have 24p in my money box in 2p coins. How many coins do I have? ☐ coins

Unit 13: Calculate mathematical statements for multiplication and division within the multiplication tables and write them using the multiplication (×), division (÷) and equals (=) signs

1. Write two number sentences to describe each array (objects arranged in rows and columns).

a) 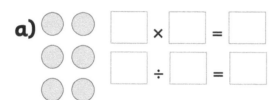   ☐ × ☐ = ☐
       ☐ ÷ ☐ = ☐

b) 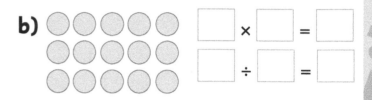   ☐ × ☐ = ☐
       ☐ ÷ ☐ = ☐

c)    ☐ × ☐ = ☐
       ☐ ÷ ☐ = ☐

2. Make an array for each multiplication, using counters. Then draw the array in the box. Use it to write a multiplication and a division sentence.

a) 5 times 2

☐ × ☐ = ☐
☐ ÷ ☐ = ☐

b) 6 times 5

☐ × ☐ = ☐
☐ ÷ ☐ = ☐

c) 3 times 10

☐ × ☐ = ☐
☐ ÷ ☐ = ☐

d) 7 times 2

☐ × ☐ = ☐
☐ ÷ ☐ = ☐

Unit 13: Calculate mathematical statements for multiplication and division within the multiplication tables and write them using the multiplication (×), division (÷) and equals (=) signs

1. Solve each addition sentence, then write it as a multiplication sentence.

a) 2 + 2 + 2 + 2 = ☐

☐ × ☐ = ☐

b) 5 + 5 + 5 + 5 + 5 + 5 = ☐

☐ × ☐ = ☐

c) 10 + 10 + 10 + 10 + 10 = ☐

☐ × ☐ = ☐

d) 5 + 5 + 5 + 5 + 5 + 5 + 5 + 5 = ☐

☐ × ☐ = ☐

2. Write a division sentence using the same numbers as the multiplication sentence.

a) 6 × 2 = 12   ☐ ÷ ☐ = ☐

b) 3 × 5 = 15   ☐ ÷ ☐ = ☐

c) 7 × 10 = 70   ☐ ÷ ☐ = ☐

d) 12 × 2 = 24   ☐ ÷ ☐ = ☐

e) 9 × 5 = 45   ☐ ÷ ☐ = ☐

f) 11 × 10 = 110   ☐ ÷ ☐ = ☐

3. Answer these questions.

a) Which has the most stickers: 9 packets with 5 stickers in each packet, or 5 packets with 10 stickers in each packet? How do you know?

_____

b) i) Danny buys 7 magazines, each costing £5. How much does he spend? Write the multiplication number sentence and calculate the cost.

_____

ii) If Sienna bought 5 books, each costing £10, who spent the most money? How do you know?

_____

Unit 14: Show that multiplication of 2 numbers can be done in any order (commutative) and division of 1 number by another cannot

1. Write a multiplication sentence to describe the arrays.

a) 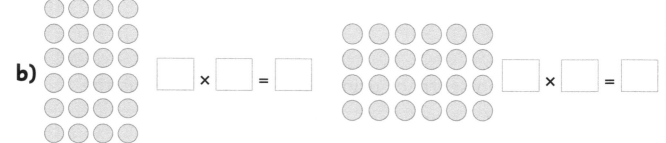 ☐ × ☐ = ☐   ☐ × ☐ = ☐

What do you notice about the multiplication sentences? _____

b) ☐ × ☐ = ☐   ☐ × ☐ = ☐

What do you notice about the multiplication sentences? _____

2. Summer places 5 books on each of 3 shelves in a bookcase. On a separate piece of paper, draw a picture that will help you work out how many books in total are on the shelves. Write a multiplication sentence that describes the picture.

_____

In another bookcase, Summer places 3 books on each of 5 shelves. How does the answer to the first bookcase problem help you to answer this problem?

_____

3. Use the three number cards to complete the four number sentences.

| 5 | 40 | 8 |

☐ × ☐ = ☐   ☐ × ☐ = ☐

☐ ÷ ☐ = ☐   ☐ ÷ ☐ = ☐

Unit 14: Show that multiplication of 2 numbers can be done in any order (commutative) and division of 1 number by another cannot

1. Complete the number sentences.

   a) $5 \times \boxed{} = 20$ **b)** $\boxed{} \times 2 = 14$ **c)** $10 \times \boxed{} = 90$ **d)** $3 \times \boxed{} = 6$

   $\boxed{} \times 5 = 20$ $2 \times \boxed{} = 14$ $\boxed{} \times 10 = 90$ $\boxed{} \times 3 = 6$

2. Draw lines to join the multiplications in the top row to the multiplications in the bottom row with the same answer.

   4×10 9×5 2×7 10×12 5×8 6×2

   7×2 2×6 10×4 5×9 12×10 8×5

3. Circle the statements that are **true**. Remember, the sign $\neq$ means 'does not equal'.

   $10 \times 5 = 5 \times 10$ $10 \div 2 = 2 \div 10$ $7 \times 2 \neq 2 \times 7$

   $8 \div 2 \neq 2 \div 8$ $11 \times 5 = 6 \times 11$ $60 \div 5 = 5 \div 60$

4. 'Trays contain 5 beads. 9 trays will contain 45 beads.'

   Tick the statements that are **true**.

   a) There will be a total of 20 beads in 4 trays. $\boxed{}$

   b) 45 beads can be shared evenly between 9 trays. $\boxed{}$

   c) 9 beads can be shared evenly between 45 trays. $\boxed{}$

   d) 5 beads can be shared evenly between 9 trays. $\boxed{}$

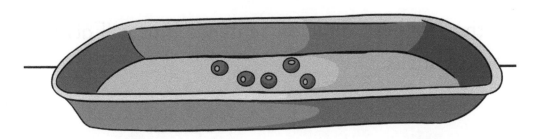

Unit 15: Solve problems involving multiplication and division, using materials, arrays, repeated addition, mental methods, and multiplication and division facts, including problems in contexts

1. Use the array of smiley faces to answer these questions.

   a) Children stand in the playground in 8 rows of 5. How many children are standing in the playground?

   ⬜ children

   b) Children are seated in the school hall in 5 rows of 8. How many children are in the hall?

   ⬜ children

   c) 40 children are split evenly on to 5 tables. How many children will there be on each table?

   ⬜ children

   d) 40 children are split evenly into 8 groups. How many children are in each group?

   ⬜ children

2. 18 shells are arranged in rows and columns.

   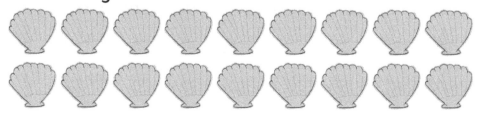

   a) Write the repeated addition sentence that will give the total number of shells.

   _____ = ⬜ shells

Unit 15: Solve problems involving multiplication and division, using materials, arrays, repeated addition, mental methods, and multiplication and division facts, including problems in contexts

b) The shells are split into groups of two shells. How many shells will be in each group? Write the repeated subtraction sentence that will give the number of groups.

_____ = 0 ☐ groups

3. Draw lines to join each problem to the number fact that can be used to solve it.

5 frogs sit on a lily pad. How many frogs will there be on 7 lily pads?	12p is shared equally between 6 children. How much does each child get?	Birds have 2 legs. How many legs do 12 birds have altogether?	5 beetles sit on each leaf of a plant. If the plant has 9 leaves, how many beetles are there?

| 12 × 2 = 24 | 5 × 7 = 35 | 12 ÷ 6 = 2 | 9 × 5 = 45 |

Unit 15: Solve problems involving multiplication and division, using materials, arrays, repeated addition, mental methods, and multiplication and division facts, including problems in contexts

1. Solve the problems using your preferred method.

a) 40 footballs are shared evenly between 10 bags. How many balls will there be in each bag? ☐ balls

b) Star stickers cost 5p each. What is the price of 7 stickers? ☐ p

c) A toaster can toast 2 pieces of bread. How many times does a toaster need to be used to toast 14 pieces of bread? ☐ times

d) Shirts have 10 buttons. How many buttons will there be on 12 shirts? ☐ buttons

e) Each flower pot holds five flowers. How many pots hold 55 flowers? ☐ pots

f) A bottle of fruit squash holds 2 litres. How much will 12 bottles hold? ☐ l

2. Complete each number problem using the numbers 2, 5 or 10, then a number from 1 to 100. Make sure it is possible to solve the problem, then solve it.

a) Ladybirds have ☐ spots. How many ladybirds have ☐ spots?

☐ spots

b) Stickers cost ☐ p each. How many stickers can you buy for ☐ p?

☐ stickers

Unit 15: Solve problems involving multiplication and division, using materials, arrays, repeated addition, mental methods, and multiplication and division facts, including problems in contexts

c) A house has ☐ windows. How many windows do ☐ houses have?

☐ windows

d) Children stand in rows of ☐.
How many rows will there be for ☐ children?

☐ rows

e) A rope is ☐ metres long. What is the total length of ☐ ropes?

☐ metres

f) A bunch of flowers has ☐ flowers. How many flowers will ☐ bunches have?

☐ flowers

Unit 16: Recognise, find, name and write fractions $\frac{1}{3}$, $\frac{1}{4}$, $\frac{2}{4}$ and $\frac{3}{4}$ of a length, shape, set of objects or quantity

1. Shade each fraction.

a) $\frac{2}{4}$
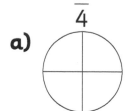

b) three quarters

c) $\frac{1}{4}$
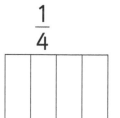

d) one third

2. Find the fraction.

a)
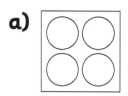

Half of ☐ is ☐.

b)

$\frac{1}{3}$ of ☐ is ☐.

c)

A quarter of ☐ is ☐.

d)

$\frac{3}{4}$ of ☐ is ☐.

e)

A third of ☐ is ☐.

f)

$\frac{2}{4}$ of ☐ is ☐.

3. Work out which fraction of each circle is shaded. Then write the fraction as a word and as a number.

a)

_____

b)

_____

c)

_____

Unit 16: Recognise, find, name and write fractions $\frac{1}{3}$, $\frac{1}{4}$, $\frac{2}{4}$ and $\frac{3}{4}$ of a length, shape, set of objects or quantity

1. Find one half of each set of objects.

   a) 6 teddy bears ☐   **b)** 12 cupcakes ☐

   c) 4 leaves ☐   **d)** 16 apples ☐

   e) 8 pencils ☐   **f)** 20 cups ☐

2. Find one quarter of each set of objects.

   a) 4 ice creams ☐   **b)** 12 books ☐

   c) 8 chairs ☐   **d)** 20 bananas ☐

   e) 16 monkeys ☐   **f)** 24 spoons ☐

3. Find one third of each length.

   a) 6 cm ☐ cm   **b)** 15 cm ☐ cm   **c)** 30 cm ☐ cm

   d) 90 cm ☐ cm   **e)** 60 cm ☐ cm   **f)** 24 cm ☐ cm

4. Answer these questions.

   a) Tom gives half of his toy car collection away. He now has 12 cars. How many did he have before? ☐ cars

   b) Leah planted some seeds. Only a third of them grew. If 5 seeds grew into flowers, how many seeds did Leah plant? ☐ seeds

   c) A quarter of the cars in a car park are red. If 5 cars are red, how many cars are parked? ☐ cars

   d) Three quarters of the pencils in a tray are green. If there are 20 pencils in the tray, how many of them are green? ☐ pencils

   e) Bryony has 40 DVDs. Is it possible to find $\frac{1}{2}$, $\frac{1}{3}$ and a $\frac{1}{4}$ of them without breaking any of them?

Unit 17: Write simple fractions, for example, $\frac{1}{2}$ of 6 = 3 and recognise the equivalence of $\frac{2}{4}$ and $\frac{1}{2}$

1. Use the arrays to complete the fraction sentences.

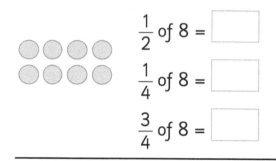

$\frac{1}{2}$ of 8 = ☐

$\frac{1}{4}$ of 8 = ☐

$\frac{3}{4}$ of 8 = ☐

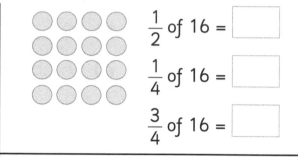

$\frac{1}{2}$ of 16 = ☐

$\frac{1}{4}$ of 16 = ☐

$\frac{3}{4}$ of 16 = ☐

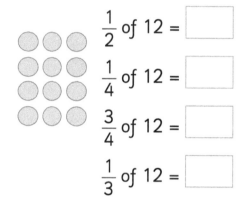

$\frac{1}{2}$ of 12 = ☐

$\frac{1}{4}$ of 12 = ☐

$\frac{3}{4}$ of 12 = ☐

$\frac{1}{3}$ of 12 = ☐

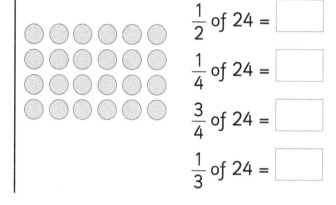

$\frac{1}{2}$ of 24 = ☐

$\frac{1}{4}$ of 24 = ☐

$\frac{3}{4}$ of 24 = ☐

$\frac{1}{3}$ of 24 = ☐

2. What fraction of each set of squares is marked with a cross?

a) | X | | | X |

fraction $\frac{\ ☐\ }{\ ☐\ }$

b)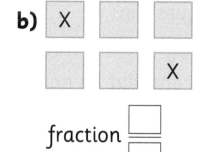

fraction $\frac{\ ☐\ }{\ ☐\ }$

c)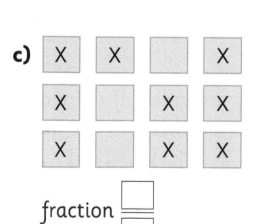

fraction $\frac{\ ☐\ }{\ ☐\ }$

d)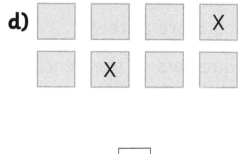

fraction $\frac{\ ☐\ }{\ ☐\ }$

Number – fractions

Unit 17: Write simple fractions, for example, $\frac{1}{2}$ of 6 = 3 and recognise the equivalence of $\frac{2}{4}$ and $\frac{1}{2}$

1. Find the fractions of the whole numbers.

a) $\frac{1}{2}$ of 10 ☐   **b)** $\frac{1}{4}$ of 8 ☐   **c)** $\frac{1}{3}$ of 9 ☐

d) $\frac{1}{4}$ of 4 ☐   **e)** $\frac{1}{3}$ of 6 ☐   **f)** $\frac{3}{4}$ of 8 ☐

2. Complete the missing numbers.

a) $\frac{1}{2}$ of 16 = ☐   **b)** $\frac{1}{4}$ of 20 = ☐   **c)** $\frac{1}{3}$ of 18 = ☐

d) $\frac{1}{4}$ of 12 = ☐   **e)** $\frac{2}{4}$ of 12 = ☐   **f)** $\frac{3}{4}$ of 12 = ☐

g) $\frac{1}{☐}$ of 8 = 2   **h)** $\frac{1}{☐}$ of 9 = 3   **i)** $\frac{1}{☐}$ of 18 = 9

3. 'I know that one quarter of a set of cars is red. If 4 cars are red, what else do I know about the cars?'

Tick the statements below that are true.

a) I know that the number of cars that are in $\frac{2}{4}$ of the set will be the same number as the cars in $\frac{1}{2}$ of the set. ☐

b) $\frac{1}{4}$ of 16 is 4. ☐

c) There will be 6 cars in half of the set. ☐

d) One third of the cars is red. ☐

4. Cross out the fraction sentences that are incorrect.

a) $\frac{1}{3}$ of 12 sheep is 3 sheep.

b) $\frac{2}{4}$ of 50 ducks is 30 ducks.

c) $\frac{3}{4}$ of 80 monkeys is 60 monkeys.

Unit 18: Choose and use appropriate standard units to estimate and measure length/height in any direction (m/cm) and mass (kg/g) to the nearest appropriate unit, using rulers and scales

1. How long is each object? Estimate first, then use a ruler to measure the length. How accurate was your estimate?

a)

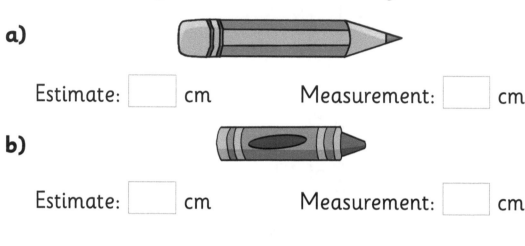

Estimate: ☐ cm   Measurement: ☐ cm

b)

Estimate: ☐ cm   Measurement: ☐ cm

2. How tall is each object? Estimate first, then use a ruler to measure the length. How accurate was your estimate?

a)

Estimate: ☐ cm   Measurement: ☐ cm

b)

Estimate: ☐ cm   Measurement: ☐ cm

Measurement

Unit 18: Choose and use appropriate standard units to estimate and measure length/height in any direction (m/cm) and mass (kg/g) to the nearest appropriate unit, using rulers and scales

3. The weighing scales have lost their pointer. Replace each pointer by drawing an arrow in the correct position.

a)

b)

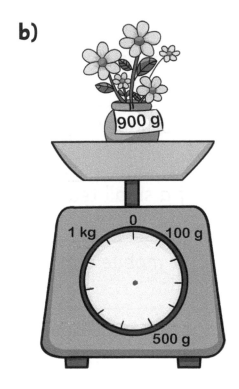

c)

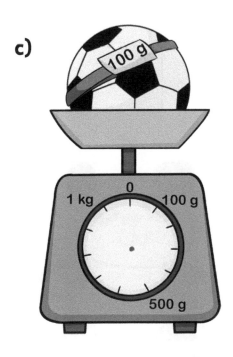

d)

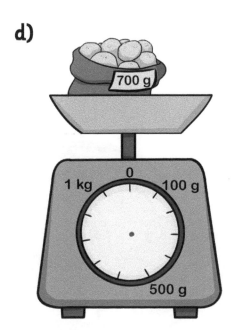

Unit 18: Choose and use appropriate standard units to estimate and measure length/height in any direction (m/cm) and mass (kg/g) to the nearest appropriate unit, using rulers and scales

1. Look at the objects to be measured. What would be the most suitable unit measurement? What equipment or tool would you use to measure the object? Complete the table.

Object	Unit of measure	Equipment/ tool
length of a paper clip	centimetre	30 cm ruler
height of a small teddy bear		
mass of a mobile phone		
height of the classroom		
length of a car		
mass of a packed suitcase		

Unit 18: Choose and use appropriate standard units to estimate and measure length/height in any direction (m/cm) and mass (kg/g) to the nearest appropriate unit, using rulers and scales

2. What is the mass of each object?

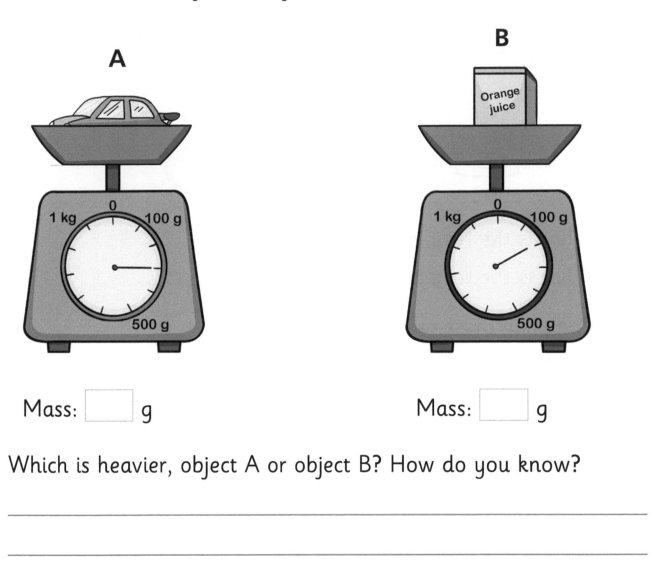

A

B

Mass: ☐ g

Mass: ☐ g

Which is heavier, object A or object B? How do you know?

_____

_____

_____

Unit 19: Choose and use appropriate standard units to estimate and measure temperature (°C) and capacity (litres/ml) to the nearest appropriate unit, using thermometers and measuring vessels

1. a)

What is the temperature of the juice? ☐ °C

What is the temperature of the tea? ☐ °C

b) Complete these sentences.

The juice is _____ than the tea.

The tea is _____ than the juice.

2. Write the measurements.

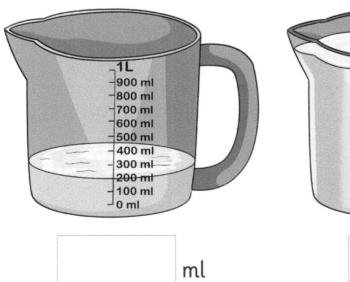

☐ ml

☐ l

Unit 19: Choose and use appropriate standard units to estimate and measure temperature (°C) and capacity (litres/ml) to the nearest appropriate unit, using thermometers and measuring vessels

1. What would be most the most suitable unit of measurement for these objects? What equipment or tool would you use to measure the object? Complete the table

Object	Unit of measure	Equipment/tool
temperature of a person		
capacity of a large bucket		
temperature of the classroom		
capacity of an eggcup		

2. What is the temperature shown on each thermometer?

a)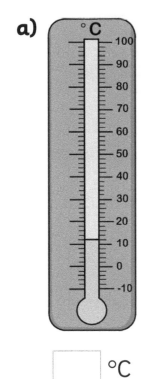

☐ °C

b)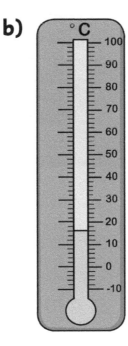

☐ °C

c)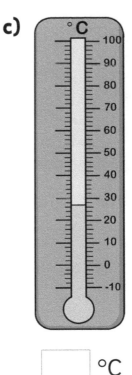

☐ °C

d)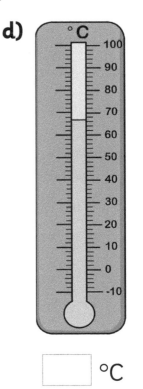

☐ °C

Unit 20: Compare and order lengths, mass, volume/ capacity and record the results using >, < and =

1. Use a ruler to measure the length of these objects.

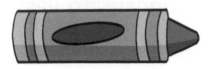

Use the information to complete these sentences.

The straw is ☐ cm and the crayon is ☐ cm.

The straw is _____ than the crayon.

The straw is ☐ cm longer than the crayon.

2. Measure the mass of a dictionary and the mass of a stapler using a set of weighing scales. Use the information to complete these sentences.

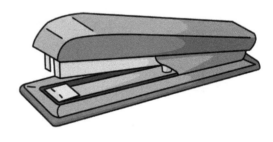

The dictionary has a mass of ☐ g and the stapler has a mass of ☐ g.

The _____ is heavier than the _____.

The mass of the _____ is ☐ g more than the mass of the _____.

3. Use > and < to compare the measures.

100 ml ☐ 100 l   115 ml ☐ 151 ml   63 ml ☐ 36 ml

Unit 20: Compare and order lengths, mass, volume/ capacity and record the results using >, < and =

1. Fill in the boxes using the symbols < or >.

a) 7 cm ☐ 9 cm   **b)** 17 m ☐ 13 m   **c)** 23 kg ☐ 32 kg

d) 74 g ☐ 47 g   **e)** 67 ml ☐ 76 ml   **f)** 48 l ☐ 39 l

g) 106 g ☐ 89 g   **h)** 99 l ☐ 108 l   **i)** 121 kg ☐ 112 kg

2. Label the number line with the approximate positions of these mass measurements.

| 75 g | 135 g | 90 g | 115 g | 60 g |

50 g   150 g

Order the measurements, from lightest to heaviest.

☐ g, ☐ g, ☐ g, ☐ g, ☐ g

3. Label the number line with the approximate positions of these capacity measurements.

| 135 ml | 85 ml | 105 ml | 65 ml | 120 ml |

50 ml   150 ml

Order the measurements, from lowest to highest capacity.

☐ ml, ☐ ml, ☐ ml, ☐ ml, ☐ ml

Unit 20: Compare and order lengths, mass, volume/ capacity and record the results using >, < and =

4. Use a ruler to measure the length of the objects pictured below.

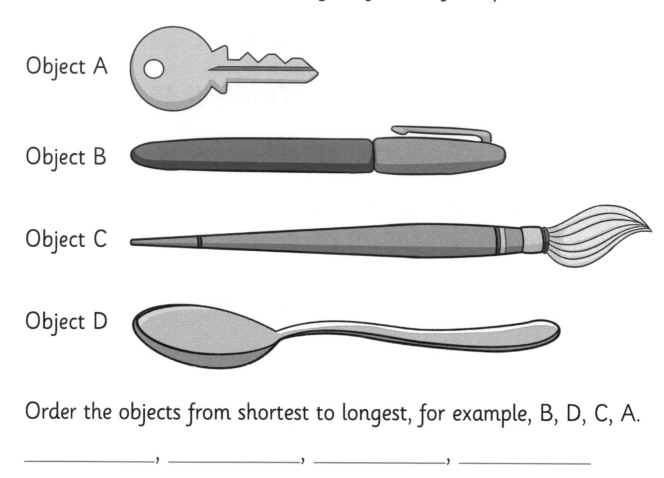

Object A

Object B

Object C

Object D

Order the objects from shortest to longest, for example, B, D, C, A.

_____, _____, _____, _____

5. Jack is measuring the capacity of three containers, A, B and C. Container B holds more than container C. Container A holds 100 ml more than container C, but 50 ml less than container B. If container A holds 150 ml, what are the capacities of containers B and C?

B [   ] ml   C [   ] ml

Order of containers, from lowest to highest capacity:

[   ]   [   ]   [   ]

Measurement

Unit 21: Recognise and use symbols for pounds (£) and pence (p); combine amounts to make a particular value

1.

FLOUR **73p**

22p

44p

BUTTER **36p**

CRISPS **57p**

SUGAR **19p**

a) Choose three items to buy from a supermarket. Draw the coins that you would use to pay for each item.

Item: _____

Item: _____

Item: _____

b) Choose another item and pay for it with the least number of coins.

Item: _____

2.

£37   £59   £13   £76   £66   1:30 £28

Choose three items to buy from an electronics shop. Draw the notes and coins that you would use to pay for each item.

Item: _____

Item: _____

Item: _____

Unit 21: Recognise and use symbols for pounds (£) and pence (p); combine amounts to make a particular value

1. Write the amounts in numbers, using the correct symbol for pounds or pence.

   a) three pounds [ ]   **b)** forty-six pence [ ]

   c) seventy-eight pence [ ]   **d)** sixteen pounds [ ]

   e) eighty-nine pence [ ]   **f)** sixty-three pounds [ ]

2. Complete the totals in the table. The first row has been done for you.

Coins/Notes	Total
one 20p coin, three 5p coins, one 1p coin	36p
four 10p coins, four 5p coins, one 2p coin, one 1p coin	
one 50p coin, three 10p coins, six 2p coins	
one £20 note, one £10 note, one £5 note, two £1 coins	
six £10 notes, three £5 notes, three £2 coins	
one £50 note, two £20 notes, seven £1 coins	

3. Jack has three silver coins in his hand. Each of the coins has a different value. List two combinations of the coins and the total of each.

   Combination 1: _____ Total [ ] p

   Combination 2: _____ Total [ ] p

4. Maisie says she can make 87p using just four coins. Is she correct? Explain how you know.

   _____

Unit 22: Find different combinations of coins that equal the same amounts of money

1. Make 57p in four different ways using these coins. You can use each coin more than once. Draw the coins you use in each purse.

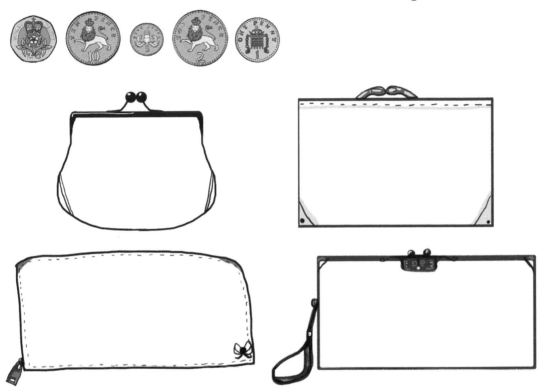

2. Find four amounts that can be made from different combinations of 5 coins, 4 coins, 3 coins and 2 coins. One has been done for you.

Amount	5 coins	4 coins	3 coins	2 coins
15p				

Unit 22: Find different combinations of coins that equal the same amounts of money

1. Find four different combinations of coins to make each amount. The first one has been done for you.

a) 28p

20p, 5p, 2p, 1p

4 × 5p, 4 × 2p

5 × 5p, 2p, 1p

2 × 10p, 5p, 2p, 1p

b) 43p

c) 37p

d) 64p

2. a) Which of these prices can you **not** make with the coins shown? The coins must make the exact amount. Circle the price tags that you cannot make.

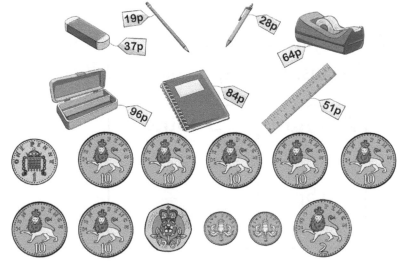

19p
37p
28p
64p
84p
51p
96p

b) Choose three items. Which coins could you use to pay for each item without needing change?

Item: _____ Coins: _____

Item: _____ Coins: _____

Item: _____ Coins: _____

Unit 23: Solve simple problems in a practical context involving addition and subtraction of money of the same unit, including giving change

1. Find the total cost of each pair of grocery items. List the coins you would use to pay for the items.

a) bananas and apples

Total cost: ☐ p   Coins to pay: _____

b) pineapples and ketchup

Total cost: ☐ p   Coins to pay: _____

c) carrots and flour

Total cost: ☐ p   Coins to pay: _____

d) flour and biscuits

Total cost: ☐ p   Coins to pay: _____

Unit 23: Solve simple problems in a practical context involving addition and subtraction of money of the same unit, including giving change

2. If you paid for each item in Question 1 with the coin shown, how much change would you get?

a)

Change: [ ] p

b)

Change: [ ] p

c)

Change: [ ] p

d)

Change: [ ] p

3. Jessica used a £1 coin to buy a packet of beads. Here is the change she was given. How much did the packet of beads cost?

[ ] p

Unit 23: Solve simple problems in a practical context involving addition and subtraction of money of the same unit, including giving change

1. Complete the additions.

a) 9p + 13p = ☐ p   **b)** 15p + 14p = ☐ p   **c)** 16p + 25p = ☐ p

d) 36p + 23p = ☐ p   **e)** 58p + 33p = ☐ p   **f)** 66p + 29p = ☐ p

2. Divit went to the shop and spent 49p. He bought at least one of each type of sticker. Which sticker did he buy two of?

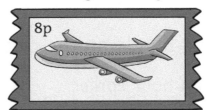

3. Complete the subtractions.

a) 10p − 7p = ☐ p   **b)** 20p − 14p = ☐ p   **c)** 20p − 6p = ☐ p

d) 50p − 28p = ☐ p   **e)** 50p − 16p = ☐ p   **f)** £1 − 44p = ☐ p

4. Noah uses two 50p coins to buy a magazine which costs 64p. He is given five coins in change. Find two possible combinations of coins he could have been given.

Combination 1: ＿＿＿＿＿＿＿   Combination 2: ＿＿＿＿＿＿＿

5. Amelia uses four £20 notes to buy a computer printer which costs £61. Find two possible combinations of coins and notes she could have been given as change.

Combination 1: ＿＿＿＿＿＿＿   Combination 2: ＿＿＿＿＿＿＿

Unit 24: Compare and sequence intervals of time

1. Order these, from the earliest time to the latest time. All the times are in the morning.

   a)

Half past 4	Quarter to 4	Quarter past 4	Half past 3

   _____ _____ _____ _____

   b)

Quarter to 7	Quarter to 8	Half past 7	Quarter past 8

   _____ _____ _____ _____

2. Draw the hands on the clocks to show the earlier and later times.

   a) half an hour earlier   **10 o'clock**   half an hour later

   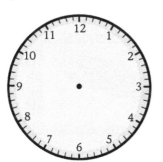  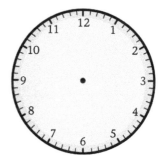

   b) quarter of an hour earlier   **twenty past 10**   quarter of an hour later

   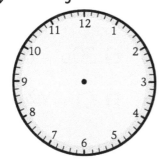  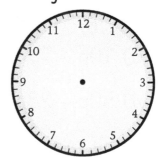

   c) 50 minutes earlier   **half past 7**   50 minutes later

   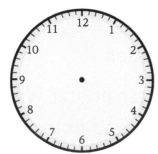  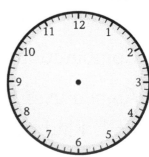

Unit 24: Compare and sequence intervals of time

1. Circle the greater amount of time.

   a) half an hour   or   25 minutes

   b) one hour   or   55 minutes

   c) quarter of an hour or 20 minutes

   d) 40 minutes or three quarters of an hour

2. Circle the clock faces that show a time between 9 o'clock and 11 o'clock.

3. Complete the table to show the difference in minutes between the two times.

Earlier time	Later time	Difference (minutes)
half past 6	quarter to 7	
quarter past 3	ten to 4	
five past 9	five to 10	
twenty past 1	half past 2	
ten to 5	half past 6	

Unit 25: Tell and write the time to five minutes, including quarter past/to the hour and draw the hands on a clock face to show these times

1. Write the time.

a)

b)

c)

_____ _____ _____

d)

e)

f)

_____ _____ _____

2. Draw the hands on the clocks.

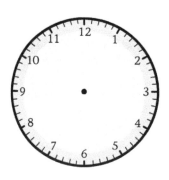

twenty past 7

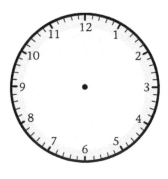

ten to 10

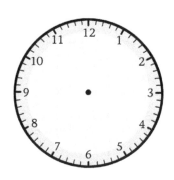

twenty-five to 3

Unit 25: Tell and write the time to five minutes, including quarter past/to the hour and draw the hands on a clock face to show these times

1. Circle the clock in each pair that shows the correct time.

a) ten past 3

b) twenty-five past 6

c) twenty to 7

d) five to 12

2. Write the times.

a) The minute hand on the clock is pointing to the 5 and the hour hand is pointing to the 3. _____

b) The minute hand on the clock is pointing to the 10 and the hour hand is pointing to the 6. _____

c) The minute hand on the clock is pointing to the 2 and the hour hand is pointing to the 10. _____

d) The minute hand on the clock is pointing to the 8 and the hour hand is pointing to the 12. _____

3. Complete the missing times.

Sienna leaves the house at ten past 8 in the morning. 25 minutes later, she arrives at the bus stop. It is now _____. Half an hour later the time is _____. This is when Sienna arrives at the cinema.

Unit 26: Know the number of minutes in an hour and the number of hours in a day

1. Complete the clock to show the five-minute intervals.

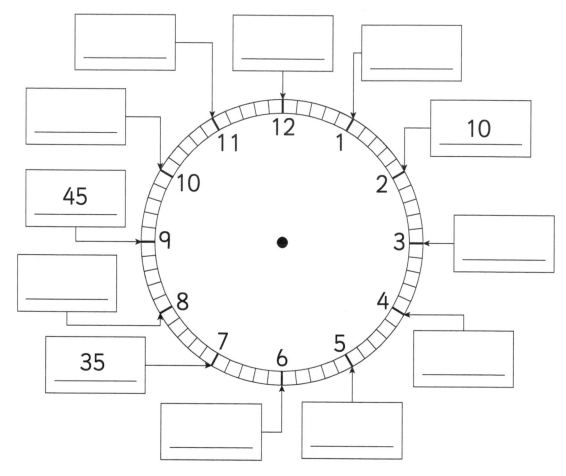

2. Draw lines to match the times in the top row with the times in the bottom row.

| half a day | half an hour | a day | an hour | quarter of an hour | 2 days | 10 days | 2 hours |

| 240 hours | 48 hours | 15 minutes | 12 hours | 120 minutes | 24 hours | 30 minutes | 60 minutes |

Unit 26: Know the number of minutes in an hour and the number of hours in a day

1. Complete the number sentences.

a) 1 hour = ☐ minutes **b)** 1 day = ☐ hours

c) half an hour = ☐ minutes **d)** half a day = ☐ hours

e) quarter of an hour = ☐ minutes **f)** 2 days = ☐ hours

g) 2 hours = ☐ minutes **h)** 5 days = ☐ hours

2. Answer these questions.

a) Chelsea begins reading her book at 7 o'clock. She stops reading at half past 7. How long did she read for? Write the time as a fraction of an hour and then in minutes. ☐ hour ☐ minutes

b) Alfie begins running a race at ten past 1. He stops running at twenty-five past 1. How long did he run for? Write the time as a fraction of an hour and then in minutes. ☐ hour ☐ minutes

c) George arrives at the restaurant at quarter past 7. He leaves at quarter past 8. How long did he stay? Write it in hours and then in minutes. ☐ hours ☐ minutes

d) Robin arrives at the football game at quarter to two. She leaves at quarter to 4. How long did she stay? Write it in hours and then in minutes. ☐ hours ☐ minutes

e) A café opens at 6 in the morning and then closes at 6 in the morning the next day. How long does the café stay open? Write it in days and then in hours. ☐ days ☐ hours

f) A supermarket opens at 12 noon and then closes at 12 midnight. How long does the supermarket stay open? Write the time as a fraction of a day and then in hours. ☐ day ☐ hours

Unit 27: Identify and describe the properties of 2-D shapes, including the number of sides, and line symmetry in a vertical line

1. Use the key to colour the shapes.

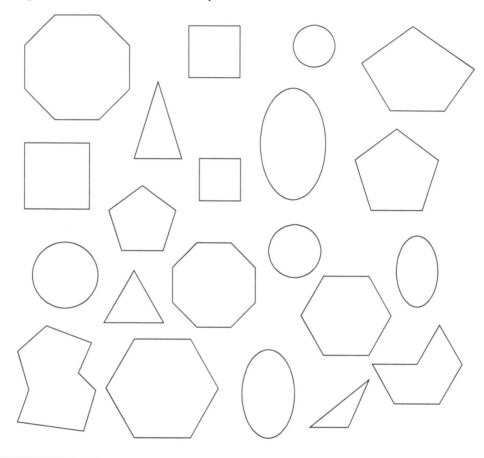

> **Key**
>
> circle – blue,   square – red,   oblong – green,   triangle – yellow,
> pentagon – orange,   hexagon – brown,   octagon – pink

2. Draw and name these shapes.

a) A shape with 4 sides, 2 of the same length and 2 of a different length. The vertices should be shaped like the corner of a book.

Name of shape: _____

Unit 27: Identify and describe the properties of 2-D shapes, including the number of sides, and line symmetry in a vertical line

b) A shape with 3 sides and 3 vertices.

Name of shape: _____

c) A shape with no sides, no vertices and 1 curved line.

Name of shape: _____

d) A shape with 5 sides.

Name of shape: _____

3. Check each shape to see if it has a vertical line of symmetry. Draw the line of symmetry.

Unit 27: Identify and describe the properties of 2-D shapes, including the number of sides, and line symmetry in a vertical line

1. Complete the table.

Shape	Number of sides	Number of vertices
circle		
	3	3
square		
oblong		
pentagon		
	6	6
octagon		

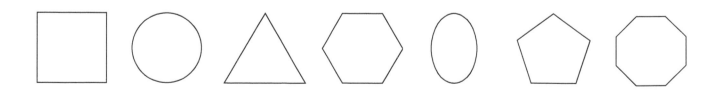

Unit 27: Identify and describe the properties of 2-D shapes, including the number of sides, and line symmetry in a vertical line

2. Complete the shapes to make them symmetrical.

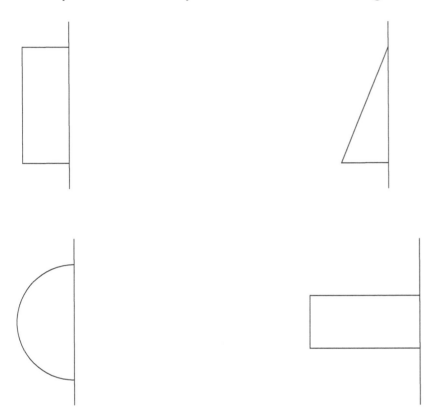

3. Circle the shapes that have the line of symmetry marked incorrectly. Draw the correct position of the line in red.

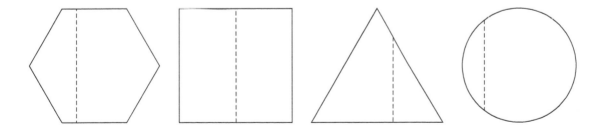

Unit 28: Identify and describe the properties of 3-D shapes, including the number of edges, vertices and faces

1. Use the key to colour the shapes.

> **Key**
>
> sphere – blue, cylinder – red, cone – green, cube – yellow,
> cuboid – orange, triangular prism – brown, square pyramid – pink

2. Draw and name these shapes. Draw real-life examples of each shape, if you wish.

a) A solid shape with identical triangular ends connected by 3 flat faces.

Name of shape: _____

Unit 28: Identify and describe the properties of 3-D shapes, including the number of edges, vertices and faces

b) A solid shape with 6 square faces, 8 vertices and 12 edges.

Name of shape: _____

c) A solid shape with no faces, vertices or edges. It has 1 curved surface.

Name of shape: _____

d) A solid shape with a square base connected to a point at the top by triangular faces.

Name of shape: _____

Unit 28: Identify and describe the properties of 3-D shapes, including the number of edges, vertices and faces

1. Complete the table.

Shape	Number of faces	Number of vertices	Number of edges
cube		8	
cuboid	6		
triangular prism			9
square pyramid		5	
sphere	0		

2. The children in Miss Bridge's class are given straws to make 3-D shapes. Name one 3-D shape that each child can make that will use all the straws.

a) Tom is given 9 straws.   Shape: _____

b) Ella is given 12 straws.   Shape: _____

c) Reeva is given 8 straws.   Shape: _____

3. I am a shape. Read my description, then guess which shape I am.

a) I have a square base and triangular sides that meet at a point.

Shape: _____

b) I have 6 faces and 8 vertices. My faces are oblongs.

Shape: _____

c) I have 2 circular bases connected by a curved surface.

Shape: _____

d) I have 5 faces and 9 edges. I have 2 triangular ends.

Shape: _____

e) I have no faces, edges or vertices. I am impossible to stack.

Shape: _____

Unit 29: Identify 2-D shapes on the surface of 3-D shapes, [for example, a circle on a cylinder and a triangle on a pyramid]

1. A solid shape was dipped in paint and a print made of one of its faces. Write the name of the shape that could have made the print. There may be more than one answer.

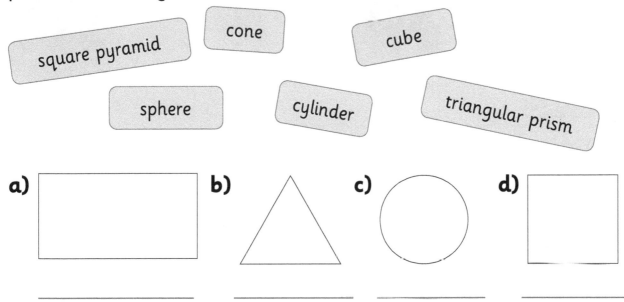

square pyramid   cone   cube

sphere   cylinder   triangular prism

a) _____ b) _____ c) _____ d) _____

2. Draw lines to match the shaded faces on the 3-D shapes to the 2-D shapes.

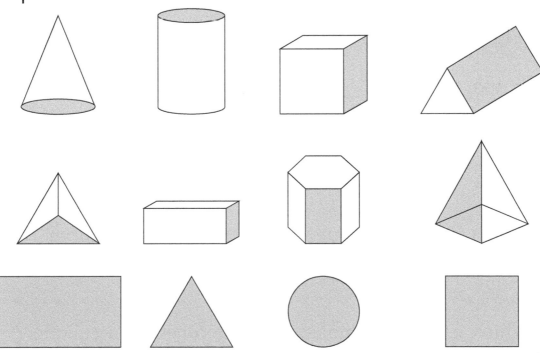

Unit 29: Identify 2-D shapes on the surface of 3-D shapes, [for example, a circle on a cylinder and a triangle on a pyramid]

1. Some children are drawing all the 2-D shapes they can find on a 3-D shape. Circle the children that have made a mistake.

| Leah draws 6 squares for a cube. | Finn draws 2 triangles and four oblongs for a triangular prism. | Taylor draws 8 oblongs for a cuboid. | Fatima draws a square and 3 triangles for a square pyramid. |

Correct any of the numbers above if they are incorrect.

2. Answer these questions.

a) Which 2-D shape makes 4 of the faces on a square pyramid?

Shape: _____

b) Which 2-D shape makes 3 of the faces on a triangular prism?

Shape: _____

c) Which 2-D shape makes 6 of the faces on a cube?

Shape: _____

d) Which 2-D shape makes 2 of the faces on a cylinder?

Shape: _____

Unit 29: Identify 2-D shapes on the surface of 3-D shapes, [for example, a circle on a cylinder and a triangle on a pyramid]

e) Which 2-D shape makes 2 of the faces on a triangular prism?

Shape: _____

f) Which 2-D shape makes 1 face on a cone?

Shape: _____

3. Pyramids and prisms are named after the shapes of their bases. Name these shapes.

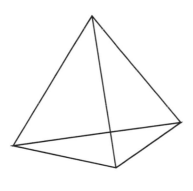

_____

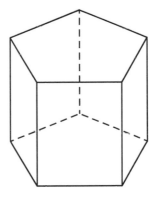

_____

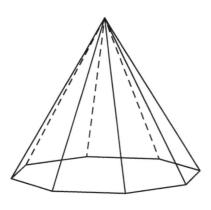

_____

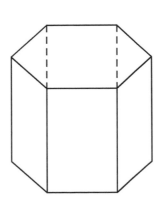

_____

Unit 30: Compare and sort common 2-D and 3-D shapes and everyday objects

1.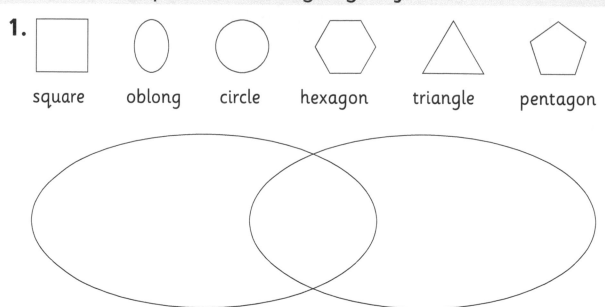

square   oblong   circle   hexagon   triangle   pentagon

Use the Venn diagram to sort the shapes. Decide on the properties you will sort and label the diagram. Draw the shapes in the correct parts.

2.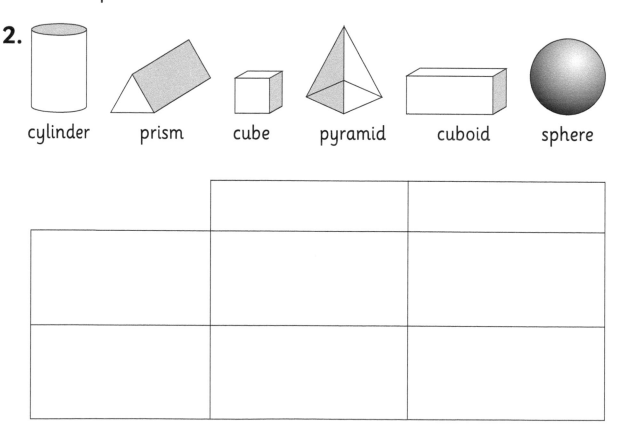

cylinder   prism   cube   pyramid   cuboid   sphere

Use the Carroll diagram to sort the shapes. Decide on the properties you will sort and label the diagram. Draw the shapes in the boxes.

Unit 30: Compare and sort common 2-D and 3-D shapes and everyday objects

1. Draw the shapes in the correct part of the Carroll diagram.

	5 vertices or more	Fewer than 5 vertices
Sides of the same length		
Sides of different length		

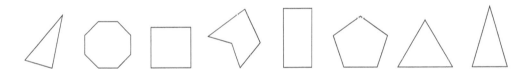

Which shape cannot be sorted? Why? _____

2. Draw each shape in the correct part of the Venn diagram.

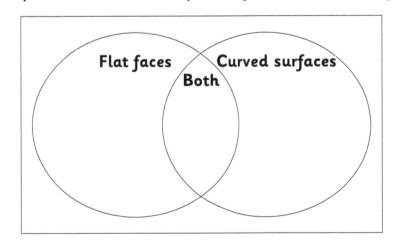

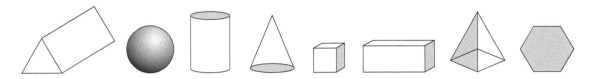

Which shape goes outside of the sets? Why? _____

Unit 31: Order and arrange combinations of mathematical objects in patterns and sequences

1. Draw the next three shapes in each sequence.

a)

b)

c)

d)

2. Create and draw a pattern using all of these shapes.

a)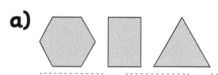

b)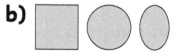

c)

d)

Unit 31: Order and arrange combinations of mathematical objects in patterns and sequences

1. Find the mistake in each pattern, then draw the correct shape in the the box.

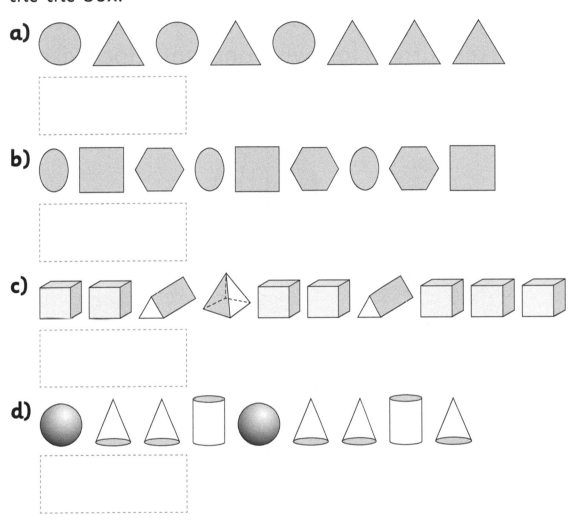

a)

b)

c)

d)

2. Answer the pattern questions.

a) Tom is making a pattern. It goes like this: hexagon, pentagon, square, hexagon, pentagon, square. What will be the 10th shape in the sequence? _____

b) Lucy is making a pattern. It goes like this: triangle, triangle, square, circle, triangle, triangle, square, circle. What will be the 15th shape in the sequence? _____

c) Ali is making a pattern. It goes like this: cube, sphere, cube, cylinder, cube, sphere, cube, cylinder. What will be the 11th shape in the sequence? _____

Unit 32: Use mathematical vocabulary to describe position, direction and movement, including movement in a straight line and distinguishing between rotation as a turn and in terms of right angles for quarter, half and three-quarter turns (clockwise and anti-clockwise)

1. Complete the sentences to describe the position of each object.

a) The hot air balloon is _____ the cat.

b) The cat is _____ the hot air balloon and _____ the dog.

c) The dog is _____ the cat.

d) The dog is to the _____ of the tree and to the _____ of the girl.

Unit 32: Use mathematical vocabulary to describe position, direction and movement, including movement in a straight line and distinguishing between rotation as a turn and in terms of right angles for quarter, half and three-quarter turns (clockwise and anti-clockwise)

2. Use the grid to answer the questions.

a) What is the position of these things?

  i) snake _____ **ii)** aeroplane _____ **iii)** lemon _____

b) Using the words **up**, **down**, **left** and **right**, write an instruction to move from one object to the other.

  i) From the house to the TV: _____

  ii) From the cup to the lemon: _____

c) Which direction are you facing?

  i) You stand by the car and face the lemon: _____

  ii) You stand by the slide and face the dog: _____

d) Using compass directions, write an instruction to move from one object to the other.

  i) From the lunchbox to the car: _____

  ii) From the dog to the slide: _____

Unit 32: Use mathematical vocabulary to describe position, direction and movement, including movement in a straight line and distinguishing between rotation as a turn and in terms of right angles for quarter, half and three-quarter turns (clockwise and anti-clockwise)

1. An aeroplane flies above a bridge. Beneath the bridge is a car. Behind the bridge is a house.

Complete the sentences to describe the position of each object.

a) The bridge is _____ the aeroplane.

b) The bridge is _____ the car.

c) The bridge is _____ of the house.

2. Here is a tortoise. Describe the turn that the tortoise makes in each question in two ways: the first, using the word **clockwise** and the second, using the word **anticlockwise**.

a) _____

_____

Unit 32: Use mathematical vocabulary to describe position, direction and movement, including movement in a straight line and distinguishing between rotation as a turn and in terms of right angles for quarter, half and three-quarter turns (clockwise and anti-clockwise)

b) _____

_____

c) _____

_____

Unit 33: Interpret and construct simple pictograms, tally charts, block diagrams and tables

1. The children in a class were asked the question, 'What is your favourite sandwich filling?' 7 children said they like ham, 12 children like cheese, 4 children like egg and 11 children like tuna.

a) Use the information to complete a tally chart, then present the data in a pictogram and a block diagram.

Filling	Tally	Number of children
ham		
cheese		
egg		
tuna		

Pictogram

☺ = 1 vote

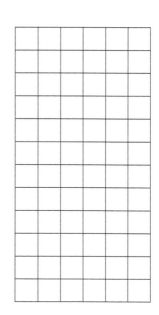

Block diagram

b) Write three things that the graphs tell you.

i) _____

ii) _____

iii) _____

Unit 33: Interpret and construct simple pictograms, tally charts, block diagrams and tables

1. Complete the missing information in the tally chart.

What is your favourite ice cream?

Flavour	Tally	Number of children
vanilla	⊬⊬ IIII	
strawberry		14
chocolate	⊬⊬ ⊬⊬ ⊬⊬ II	
cookie dough		16
bubblegum	⊬⊬ ⊬⊬ II	

How many children prefer chocolate? ☐

How many children prefer bubblegum? ☐

2. Use the table to complete the information in the pictogram.

Favourite fruit	Number
apple	4
strawberry	8
banana	12
orange	10
melon	6

☺ = 2 votes

apple	☺ ☺
strawberry	
banana	
orange	
melon	

Unit 34: Ask and answer simple questions by counting the number of objects in each category and sorting the categories by quantity

1. Tom lives on a farm. He drew a picture of the animals that are kept on the farm.

a) Complete a tally chart to show how many animals there are of each kind.

Animal	Tally	Number
goat		
horse		
chicken		
sheep		
cow		

b) Put the data from the tally chart in a block diagram. Then use the diagram to answer the questions.

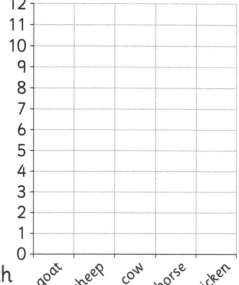

i) How many chickens are there? ☐

ii) How many sheep are there? ☐

iii) Which animal is present in the greatest number? _____

iv) Which animal is present in the fewest number? _____

v) Out of cows, chickens and goats, which animal is most common? _____

Unit 34: Ask and answer simple questions by counting the number of objects in each category and sorting the categories by quantity

1. The pictogram shows the number of doughnuts sold at a supermarket from Monday to Friday.

Number of doughnuts bought

Monday	⊙ ⊙ ⊙ ⊙ ⊙ ⊙
Tuesday	⊙ ⊙ ⊙ ⊙ ◡
Wednesday	⊙ ⊙ ⊙ ◡
Thursday	⊙ ⊙ ⊙ ⊙ ⊙ ⊙
Friday	⊙ ⊙ ⊙ ⊙ ⊙ ⊙ ⊙ ◡

⊙ = 10 doughnuts

a) Are these statements true or false? Tick the boxes.

Statement	True	False
On Tuesday, 45 doughnuts were sold.		
On Thursday, 70 doughnuts were sold.		
Fewer doughnuts were sold on Tuesday than Wednesday.		
More doughnuts were sold on Monday than Wednesday.		
The same number of doughnuts were sold on Monday and Thursday.		
Most doughnuts were sold on Monday.		

b) Write three more true statements using the information in the pictogram.

_____

_____

_____

Unit 35: Ask-and-answer questions about totalling and comparing categorical data

1. Complete this activity.

a) Use a spinner marked A–E. Spin it 50 times and use the tally chart to record the letters spun. Complete the Total column.

Letter	Tally	Total
A		
B		
C		
D		
E		

b) Construct a block diagram to represent the results. Each block stands for 2 spins.

c) Use the block diagram to answer these questions.

i) How many As and Bs were spun altogether?

ii) How many Ds and Es were spun altogether?

iii) How many more/fewer Cs were spun than Ds?

iv) How many more/fewer As were spun than Bs?

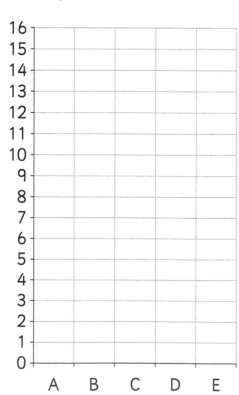

Unit 35: Ask-and-answer questions about totalling and comparing categorical data

1. The children in a class were asked the question, 'Which way do you prefer to fasten your shoes: with stick-on strips, laces, a zip or do you prefer slip-on shoes?' Use the block diagram to answer the following questions.

a) What is the total number of children that prefer shoes with stick-on strips or laces?

b) What is the total number of children that prefer shoes that are slip-on or have a zip?

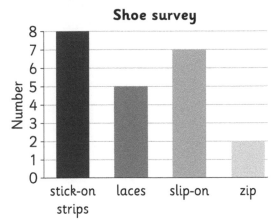

Shoe survey

Number

8
7
6
5
4
3
2
1
0

stick-on strips   laces   slip-on   zip

c) How many more children prefer shoes that fasten with stick-on strips than a zip?

d) How many fewer children prefer shoes that have laces than are slip-on?

2. The data for the monthly sales of a car company is presented in a pictogram. Use the pictogram to answer the following questions.

a) How many cars were sold in February and March?

b) How many cars were sold in April and May?

c) How many more cars were sold in April than in February?

d) How many fewer cars were sold in May than March?

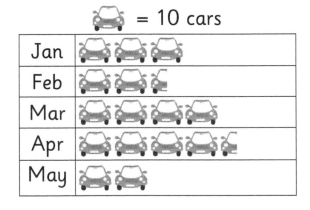

= 10 cars

Jan	
Feb	
Mar	
Apr	
May	

Notes